"

"*The Science of Story* is a life companion, one of those books that when you read it, you can never go back to previous thought patterns. Self-discovery, life-discovery and true awakening reside within the pages of this book. Delightfully delivered in a clear and concise style. When you truly understand the science behind story you will understand, at a deeper level, how life works. *You can hide from it, or embrace it, it's your choice.* Adrea gifts us the opportunity to learn the answers our hearts and minds desire to comprehend. Don't deprive yourself from YOUR story!"

KAREN MCDERMOTT, PUBLISHER

Copyright © Adrea L. Peters

First published in 2020
by MMH Press
Waikiki, WA 6169

All rights reserved. No part of this book may be used or reproduced by any means, graphic, electronic, or mechanical, including photocopying, recording, taping or by any information storage retrieval system without the written permission of the copyright owner except in the case of brief quotations embodied in critical articles and reviews. Although the author and publisher have made every effort to ensure that the information in this book was correct at press time, the author and publisher do not assume and hereby disclaim any liability to any party for any loss, damage, or disruption caused by errors or omissions, whether such errors or omissions result from negligence, accident, or any other cause. This book is not intended as a substitute for the medical advice of physicians. The reader should regularly consult a physician in matters relating to his/her health and particularly with respect to any symptoms that may require diagnosis or medical attention.

Interior and cover design: Cassandra Neece

National Library of Australia Catalogue-in-Publication data:
The Science of Story/Adrea L. Peters

ISBN: (hc) 978-0-6488499-9-5
ISBN: (sc) 978-0-6488203-5-2
ISBN: (j) 978-0-6488203-4-5
ISBN: (e) 978-0-6489376-3-0

The Science of Story

Mastering Your Nature

ADREA L. PETERS

DESIGN BY CASSANDRA NEECE

*For my grandmothers,
Katie and Thea
I love you.*

Table of Contents

Foreword from Amber Lilyestrom · 9

The Principle of Inclusions · 11

The Matter of Mindset · 16

The Electricity of It All: WHO · 36

Vibration = Voice · 56

The Principle of Reason: WHY · 73

Let's Postulate: THEME · 87

Let's Classify: GENRE · 99

The Guts of Story: WHAT, HOW, WHEN and WHERE · 113

Transition Time: THE END · 131

Your Story Lab · 138

Vetting a Story · 149

From Here Forward · 178

Acknowledgments · 180

About the Author · 182

About the Designer · 184

Foreword

When my daughter was just three years old, I kissed my family and flew 5,000 miles from home to attend my very first writer's retreat. As a self-proclaimed homebody, the decision to make this journey was exceptionally out of character, but something whispered to me - I knew I needed to get on that plane.

As soon as the plane landed, we emerged into the warm Hawaiian air. Tenderhearted and shaky, I walked to baggage claim looking and was immediately greeted with warm hugs and a beautiful flower lei. My beloved friend and guide, Adrea had orchestrated this divine gathering on the holy lands of Kauai. We were there to spend the week in cabins on the beach with a group of women who had come from far and wide to meet themselves.

In our sessions, we cracked into the mechanics and magic of story. We wrote and walked on the sand. We laughed and ate gorgeous food together. We opened the doorway to new parts of ourselves we had long ago locked away. It was heavenly. Little did I know, those few days spent in paradise further from home than I had ever been before, would be the spark that lead me back to myself.

I learned that my story was here for me, not only to share; but, most of all, to bring me closer to the truth of my own living. The heartful and evolutionary way Adrea brings story to us is about so much more than finding the right words, it's about coming back together and freeing ourselves from the inside out.

The Science of Story is a treasure. The exercises are a pathway to deeper clarity and true freedom. This book cuts through the noise and gets to the only truth that matters...yours. I am so giddy with excitement about this book and getting to share it with my students. It is a gift that will give for as long as we are here and human.

> *"The more you understand and encourage your Story, the sweeter every aspect of your life becomes." - Adrea Peters*

What a miracle that we get to be the heroes of our lives! Our story can be the chains that bind us or the key to set us on a path of true freedom and fulfillment. Thank you, Adrea, for getting to the heartbeat and walking with us on the pathway to sharing who we really are in this lifetime through our magnificent and magical stories.

xo

amber

Amber Lilyestrom
Branding Strategist and Business Mentor
Host of the Amber Lilyestrom Show Podcast

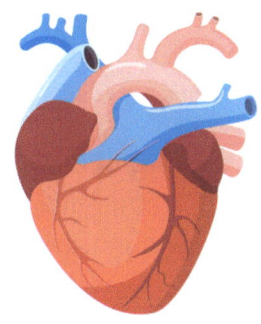

The Principle of Inclusions

It is my pleasure to have you here in *The Science of Story*. I aspire to create a beautiful, welcoming space for you in these pages. I hope you will enjoy it as much as I have loved piecing it together over the years. It will evolve, as everything does, naturally and forevermore. Maybe you'll be here for a short visit, or perhaps you and I are bound for life. Either way, I welcome you completely. We're going to have such a great time!

Story has become a bit of a buzz word which I could not be happier about!

It's a **dream come true** for me that people are ready to embrace Story.

I'm now finding there's all sorts of advice bouncing around demanding we get our Stories straight in order to succeed. **But what exactly *is* story?**

It's been a bit reserved for writerly types, hmm? Let's quash that antiquation here and now forevermore. Everyone is a storyteller. You are the novelist, playwright, screenwriter, poet, lyricist, troubadour, and raconteur of Your. Life. Story. Yet at the moment few of us know what that actually means. My goal is to change that.

Like arborists teaches us about trees and cosmologists offer knowledge of the stars, I am a Storyologist anxious to share the innerworkings of Story with you.

We tend to watch, read, mimic and envy the stories of *others* more than our own. This is good because we learn through comparison. Research, including eavesdropping and scrutinizing, invites our ideas to come forth.

However, once we see ourselves as the Hero/Heroine of our life rather than giving that role to another, **we *become* the person we dream of being.**

It can be scary. I get that. It doesn't worry me in the least. I believe in you. Always. How? Story. Triumph is what heroes do. That includes You. That IS you.

In geology there is a wonderful term, the *Principle of Inclusions,* that, in essence, explains that rocks are made up of older rocks. Striations highlight the layers. **Inclusions are distinct, but not separate. Inclusions make rocks, *and us,* beautiful, interesting, and unique.**

Have you heard of *Kintsugi,* the Japanese art of "golden joinery" where breaks in ceramics are treated as part of the object's history and beauty? Inclusion as art. That. Is. Story.

We are the sum of our whole experience.

The Principles of Inclusion

Fighting the inherent nature of inclusion is fruitless and defeating. Where exactly do we think our history goes when we leave it out? Trying to contort around who we are as opposed to including ourselves in our story creates unnecessary struggles, obstacles and sabotage. It's such a tough path to carve... until you understand Story.

How you include it is completely up to you.

Suddenly heartache is a good thing. Betrayals, lies and secrets are appreciated because they helped you become you. There is nothing to fear when you recognize the cadence of Story.

Living the story of *your life* is as good as it gets.

And that is exactly what you will be given the tools to do in The Science of Story. By mastering your nature: how you tick, what you love, what matters most to you, and how you express everything, you will begin to experience a fortitude you may never have dreamed possible. And you will, quite possibly, have more love flowing from you that ever before. I hope that excites and inspires you enough that you devour the material in this book, page by page, sentence by sentence, word by word. I hope the light within these pages surrounds you as I intend, with love, with grace, and with absolute knowing that your story matters., This book is for you. It is of you. It is yours.

A few things of note before we begin, please. Thank you.

One
This book is short, by intention. It was originally drafted for my students, clients, and attendees of writing retreats and workshops where I share my insights on Story. As I sat down to share it in proper book form, I realized it needed to remain short with as

few examples as possible. Story is not about me. It's about you.

Two
This book is intended to be your companion. I hope it is your partner as you bring your dreams to life. If I'm doing my part, my exposition will trigger your imagination. When your start to drift into your own story, and ideas comes to you: Stop reading and drop your pen to the page or fingers to the keyboards and create. Thank you.

Three
You will see icons used throughout this book to represent pieces and parts of Story. Here are their definitions:

 =**Idea**=*What Envisioned*

 =**Today**=*Who/Theme*

 =**The Sign**=*When and Where*

 =**Resist to Persist**=*Why/Theme*

 =**Help Arrives**=*How*

 =**All In**=*How/Theme*

 =**Idea**=*What Actualized*

Four

All promising equations (the math version of Story) have proofs. They are assumptions that logically guarantee a conclusion. Proofs are the Known bits (themes) that are solid as we enter the Unknown (ideas) bits. They help us feel safe as we enter the darkness. These Proofs are a blanket I drift over your shoulders to let you know all is well. Please relax and enjoy.

Proof 1: *We are the stories we tell.*

Proof 2: *We are what we create.*

Proof 3: *We are our ideal client/reader/partner/friend/soulmate.*

Proof 4: *We find what we seek.*

Proof 5: *Story defines experience.*

Proof 6: *Triumph is the only outcome, ever.*

Proof 7: *You are loved.*

Proof 8: *You are the Hero/Heroine of your life.*

Thank you for being here.

My gratitude for your presence is immeasurable.

The Matter of Mindset

Communication, in the form of written words, spoken words, silent thoughts, loud screams, points of view, sharing, listening, artistic creations, personal style, and body movement, ticks and sensations, are based on the same thing: STORY.

Story is the biology of human interaction. It's the sugar in the cookie, and the salt, butter and flour. The pulse in the blood. The water in the ocean, and the current, creatures, and debris. The spark in the soul. The light in the tunnel, and the dark.

Story is a Big Deal. Understanding it wields the power to enrich us beyond measure.

Stories connect e v e r y t h i n g.

No Story=No Way to Connect

When our stories synch or resonate with someone else's, we are forever connected.

This may be as simple as being the same age. Or born in the same country. Or went to the same high school. It may go much deeper, such as a being a mother, or survived a similar traumatic event. A car accident or divorce, or the unexpected death of a someone we love. When our story shares the same traits of another's, we connect. We link. We intertwine. And from there, we grow.

No Connection=No Point.

Story is the construct that links e v e r y t h i n g. But if there is no shared story, there is little we can link to. We simply cannot seem to Connect. We don't agree. We can't see the other's point of view. This may never happen to you. Or it may happen all the time. It all depends on the story you tell.

Our stories don't simply *matter*. They ARE matter. Our stories are the "stuff" of us. They are the materials that form us, and they are crafted from our Mindset. Our minds evoke our feelings, and our feelings bring forth more life from within us. Our feelings are alive. They are the energy of us. The fuel. The electricity.

We are ticking time bombs or smooth sailors, depending on our Mindset.

There is no end to the stories we create. With time and training, we can kind of manage some of them. But mostly they are old, and habitual, and forgotten, like a TV left on in the basement you rarely spend time in.

The bit that gets lost, and holds us back, is that you are always the one in charge. You can walk down to the basement (read: your past) and turn off the TV. The trick is you have to walk out of the basement and go back upstairs (read: stay present). It's not the easiest task, I know. The good news is that it is only Mindset. And we can mold, manage, hone, train, love and adjust it all in a thought or two. It's all up to you.

You really are the **only** decision maker in your story. That can be daunting or delightful. You decide. Doing nothing is also a decision. And may be the story of your life, thus far. If that's true for you, it gives me shivers to think that you were brave enough to join me in this journey! Well. Done.

I want you to know that you already possess all you need. You are your best example.

Know you, know everyone.

Let's dig into our stories, shall we? (I'm hearing shouts of **raucous** yeses and wahoo's! Excellent. Thank you.)

Mindset Exercise No. 1: Defining Creating

The **FIRST Mindset Exercise** we have requires a pen. If you are more comfy typing, then please grab your computer, typewriter or tablet for this.

Once you have that, please share with me, **writing in third person,** what creating is.

Remember that first person means the subject/pronoun is "I". First person written on this topic would be, "I think creating is extraordinary. When I create, I am filled with a sense of profound freedom. I like to be alone with my keyboard typing whatever I think and feel for hours on end."

Second person uses, "you" as the subject. For example, "You may think writing is so hard and something only *writers* do. You may also find it irritating to hear that anyone can write. Yet, you can write. You write emails and texts, posts and papers. Now is your time to write your story. You matter. Your story matters. You have what you need you magnificent storyteller you!"

Third person POV (point of view)—the assignment for this exercise—is written with the pronoun/subject of he/she. For instance, "Bianca is a prolific jeweler. She believes creating is a window into many worlds, from those that have come from centuries before to those yet to be discovered. Whilst creating her art for you, she never ceases to find it feeds her life force. Creating is not easy. She struggles. But she knows that is part of her jewelry getting better. She understands that her love for sharing what she creates is all that matters and is so fulfilled when one of her pieces sells."

It's time. Inhale, please. Hold it there and close your eyes. Exhale. One more time.

The Science of Story

Inhale. Close your eyes. And exhale. Begin. Write for at least five minutes without stopping on the topic of WHAT CREATING IS. There is no time limit, keep going until you have nothing left. Write to exhaustion. Go everywhere, be everywhere.

Creating is . . .

Please stay in third person. Thankyousomuch!

The Matter of Mindset

The Science of Story

This is an excellent exercise to return to when you doubt yourself and your abilities.

Mindset Exercise No. 2: Be Gone Distractions! Be Gone!

One thing we creators have in common is our absolutely jaw-dropping talent for *not* creating. We will do almost anything—yes, even pluck nose hairs—to avoid the keyboard or the newsletter or office chair. We must overcome this. No. I promise I am not asking you to do impossible. We can learn to tether this talent for procrastination and avoidance in order to funnel it to our creations. I n c l u d e them. For instance, when I am writing a novel, and my nose itches, my character's nose itches! Then it's gone. I've done something with it, and it is free. And so am I. To create more.

Therefore, I ask you to keep a list of all the ways you avoid creating. Let me give you some examples: doing the dishes, texting, reorganizing cupboards and drawers, laundry, picking fights with your children, painting nails, "researching" on Google or Facebook, social media scrolling, shopping, phoning a friend, emailing anyone you've ever known, petting the cat or dog, washing the dog, washing the car, cooking, flossing, shelving books by color. And that's just in my last hour. I'm teasing, but it is real. I'm inviting you to Shift. And Funnel your distractions into your creations. People will resonate with you more than ever before. You are them and they are you.

Your distractions are as shared as your triumphs. (Best news ever!)

Until you easily catch yourself and begin to include your distractions into your everything, you may want to keep this list going.

The Science of Story

Things that Distract YOU (and that'll NO longer stop you from writing your story):

The Matter of Mindset

May you learn your go-to moves and gently nudge
yourself back to creating.

Mindset Exercise No. 3: Obsessions/Passions

Ready for the THIRD Mindset Exercise? Fantastic! This is my favorite. I love it. I can't wait for you to do this one!

Our obsessions/passions are our best friends. You may not want this to be true. You may want to revise your list. Please don't. Own your obsessions. They tell you everything you need to know about yourself. You don't need to know why you are obsessed with what you are obsessed with. That will come with time, and understanding, of story.

Ready. Set. Write everything that you are OBSESSED with! Think PASSION. Think LOVE. Think CAN'T GET ENOUGH OF. Don't overthink. Instead, write down whatever you OVERTHINK! That's a passion/obsession. And a key component to every one of your stories.

Please use the table to capture yours.

Obsessions/Passions!

The Science of Story

When we honor what we are passionate about,
everything falls into place.

Mindset Exercise No. 4: Your Ideal & Breakthrough Traits

I plead and beg you to do this exercise. This is what I consider a life-altering exercise. It will give you sacred insight into yourself, so I hope you take advantage of it.

Please take your time. There is no time limit. You can erase, modify, ponder and then accept. No one has to see your list. Yet, I would ask you never to be ashamed or embarrassed about who you are. You are beloved. You are extraordinary. You. Are. Everything. & Everything. Is. You.

Part One: Write the eight to ten words or short phrases that are your **IDEAL qualities**. The things you want to be, you long to and sometimes have described yourself as. Write the BEST OF YOU!

The Science of Story

IDEAL TRAITS! GO!

Mine are: HILARITY—TRUST—HONESTY—LOVE—CONNECTION—IMPECCABLE—FREE—RICH--SMART

The Matter of Mindset

Well done! Ready for Part Two? Terrific!

Part Two: Take the words/phrases that ARE YOU and write your OPPOSITE to those words down on the page. **These are your BREAKTHROUGH traits.**

Please remember that there will be many opposites to your IDEAL traits. For instance, take my trait of HILARITY. The opposite of that might be: DEPRESSED, SHAME, CRYING, MEAN, BITTER, ANGRY, HARSH, NOTHINGNESS, SILENCE, SADNESS and so on. I have found this part is rather easy, so please let the opposites flow. You are probably quite familiar with them. (I say that because it is how story works. No judgement. Simply the truth of story, of life, of nature, of everything.)

Ready? Set? Go! Breakthrough/Opposites onto the page please!

The Science of Story

BREAKTHROUGH TRAITS! GO!

Mine are: SADNESS—DOUBT—LIES—DETACHMENT—SEPARATION—NEGLECT—CHAINED—LAZY--UNDERESTIMATED

Welcome Back!

How are you feeling? Wobbly? Stunned? Chill? Revealed? Floating? So good! Well! Done!

When I did mine, I was shocked. They resonated so severely, I burst into tears. I am not someone who bursts into tears. I can name the other two times, okay? When my first dog, Prince, was hit by a car and subsequently left his body and when the 20-something son of my dear friend, Nancy, told me that she was not going to live and there was no more time to say goodbye.

Here is why we do both IDEAL and BREAKTHROUGH: the space between the first set of your words and their opposites, represents your, or your character's, or client's, journey(s).

They represent you in this life. They direct your purpose. They reveal your meaning. They underlay your every-day-on-the-planet-things-to-be-ness.

Talk about insight!

Talk about powerful!

This gives you everything you need to get serious about the business of being you.

And once you master you, there is no limitation, and no constraints to your offerings. Nope. None. Only more. Always m o r e.

The Science of Story

A full story=Beginning+Middle+End
It is simply the journey from a Breakthrough Trait(s) to and Ideal Trait(s).

Before we move forward, I want this to sink in a bit more. You could actually stop here and not read forward. All stories are from a to b to a. Circler. Revolving. Evolutions. The more intrepid the journey the more letters of the alphabet applied. The base always remains the same. Ideal to Breakthrough to Ideal. Often the Ideal is not realized, or only experienced in brief bursts until the Breakthrough is broken through and you are you more than ever before.

Let's look at a common Ideal: HAPPINESS. And let's say the opposite/Breakthrough is FEAR.

If I were writing a story about someone seeking happiness, I would plot events that show the following emotional journey.[1] This is not the only set of emotions that bridge the gap. There are countless choices. I simply want to provide an example so you can do this on your own.

FEAR gives way to → **Panic** which gives way to → **Envy** which gives way to → **Annoyed** which gives way → **Self-Pity** which gives way → **Overthinking** which gives way → **Uncertainty** which gives way → **Impatience** which gives way → **Defeat** which gives way → **Exhaustion** which gives way → **Release** which gives way → **Hope** which gives way → **Positivity** which gives way → **HAPPINESS**

Remember how I mentioned our feelings being the fuel. See that here? I hope so!

The great news is, and there is a lot of wonderful in every journey, once you travel the full journey, you don't have to repeat every step every time. You don't always swing all the way from FEAR–to–HAPPINESS. You hit Impatience, realize you're actually Exhausted, then slow down/have a cat nap to Let Go/Release then wake up feeling Hopeful and Positive and viola, Happy in about 10 minutes.

Our pairs create a balanced life. Emotional homeostasis is on offer here. Can you imagine anything better?

You can use this tool for the characters of your novel, for your clients, for your work and love partners, for your businesses and products, and on and on. It's meant to be simple. It's meant to be easy to remember. Our dreams are meant to be attained, not perpetually chased.

Please begin to shift your thinking to opposites. Moon/Sun. Wet/Dry. Day/Night. Impeccable/Neglected. Wonder/Closed Off. Lost/Found. Seen/Hidden. Peace/Chaos. Both sides are needed, or we would cease to have purpose, meaning, or story in our lives.

Please do not worry if you feel lost or confused or distracted. Maybe you want to put this book down (or through it away). That is PERFECT. And to be expected. We are entering the second stage of story: No Way! No How! Resistance is critical to life/story. I like to call it Discernment. That evolves into Wisdom. You are right on track. Always.

In the next chapters, we will put all this into place so please hold tight. In fact, speaking of holding tight, please wrap those gorgeous arms of yours around yourself and squeeze. You just changed your life for the better. And learned your first set of Story Tools, you master you. Way. To. Go!

---

[1] Tools used to map this: Abraham-Hicks Emotional Scale and the Emotional Wound Thesaurus: A Writer's Guide to Psychological Trauma by Angela Ackerman and Becca Puglisi.

The Electricity of it All: *WHO*

To keep on with the opposite paradigm, we all have two poles—North and South—like our great mama Earth. And what makes you move/rotate is magnetism. They are attracted to each other, so they are converging. It is as if they are meant for each other and their energy, surging toward each other, creates a magnetic field which gives way to electricity/usable energy.

In terms of Story:

North=Ideal Traits
South=Breakthrough Traits
Magnetic Field=You
Electricity/Usable Energy=Style
Vibration/Energy Used=Expression of Style=Voice

The most important facet of the "who" of you is your Character. Character is expressed in Style and Voice. This is nothing new. You learned this years ago, yes?

I use the term Style because it is such an inclusive word. Tone, cadence, look, movement, likes/dislikes, where you live, what you drive, who you hang out with, obsessions, habits, beliefs... ALL=style. Most wonderfully put:

STYLE is the ART of You.

When your Style matches another's style—hopefully friends and clients—it's awesome and fun. And when it doesn't, interacting (inter-styling) isn't as fun, but is it necessary and enlightening! We cannot be without others. How would know who we are? To compare is to grow.

Remembering that when our Ideals include Freedom, Love, Well-Being, Happiness, all Positive traits, we have to have exposure to the opposites to get there. It is the only way through. That is the "law" of story. Breakthrough to Ideal. Opposites/comparisons give us clarity and celebration of our Ideals.

We compare/judge to gain insight. Both sides (North and South) are required. They create a magnetic field filled with electricity/energy, remember? Compare to discover. Our whole goal as Scientists of Story is figure out how things work or don't work for us so we can evolve our stories and invite our lives to become more. Always more.

Your Supporting Cast

Speaking of others... let's get to know how our relationships fuel us. The people in our life, past, future and present, are essential contributors to WHO we are. And we are, I promise, the better for it. Here is the golden rule on others, my dear friends, they are here for us. We grow amongst others, in spite of others, because of others, and without the others we wish were in our life. To be human, is to connect, via Story. Our relationships give us fortitude, as in bravery, courage, strength, spirit, guts, grit, resilience and resolve to evolve. Boy do I love the word fortitude. Whew!

We become more because of others.

Therefore, it's a wonderful thing to map out our Supporting Cast. I drafted an infographic to show you how we relate to the Cast of Characters in our story (s) and Cass made it absolutely astonishing. Through this visual it becomes clear how many folks influence our lives. You know this stuff. You simply may not have mapped it out. I know I hadn't! In a few pages, there will be a few beautiful blank versions for you to map yours.

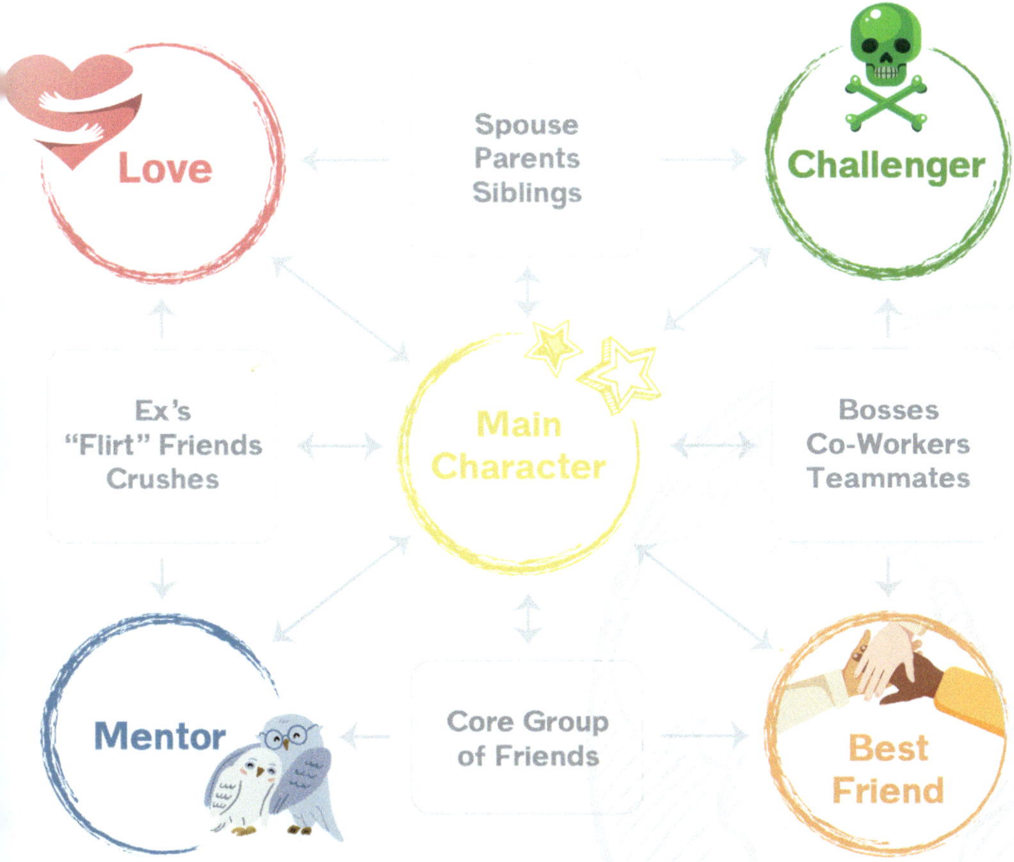

The Electricity of it All: WHO

Love
Spouse
Parents
Siblings

Challenger

Main Character

Ex's
"Flirt" Friends
Crushes

Bosses
Co-Workers
Teammates

Mentor

Core Group
of Friends

Best Friend

39

1. **Main Character (YOU)** are in the middle because are the center of your life, remember? I know, we do forget sometimes, don't we?

2. **The Challenger** is your primary Opposition—and this is a person, not you. It's someone who really pushes our buttons. A mirror you don't want to look into.

3. **The Best Friend** is your go to for support. They are Truth Tellers that sometimes sting and singe us. And no matter what, you always reunite. This can be also be a work partner or teammate.

4. **The Mentor** is the one that will take you to the next level. Could be a teacher. A boss. A publisher. A parent or grandparent. They are in your life to help, which sometimes is a kick in the bum. Unlike the Best Friend, they are not always in your life for more than a little while. They offer wisdom and guidance in whatever way will get through to you, then they move on. They are not necessarily impacted by you as much as you are impacted by them. More on this role in The Guts of Story chapter.

5. **The Love** is where your heart is, again, in the form of a person. A mirror you want to look into.

These cast members change Story to Story. Your home life story may have a different Cast than your career story. Your dream life story, your family story… your trauma story, all, or at least some are different players. and so on.

In between quadrants are what I call "sub" characters. They certainly are not "sub" but when we classify, as all good scientists do, that's what they'd be dubbed so that's where this brain went! These people in our life are essential and usually remain for long periods of time, if not the entire span of our life, so they are far from sub. They may dart in and out, influencing us greatly, but are not necessarily primary in our daily life. Again, it depends on the Story your telling. I simply want you to see how many relationships we have active in each and every story because it helps us understand the science of story, meaning how story works.

Let's go through the Sub Cast, shall we?

1. **Spouse/Parents/Siblings** fall between Love and Challenger. And are they not that? We based our love stories on our parents. And our siblings are often our biggest challengers! As are our parents and spouses. See how this works? Love it. I truly do. We may not want to admit the patterns we repeat, but when we see this, we see it's a bit unavoidable. May as well roll with it.

2. **Bosses/Co-Workers/Teammates** are between Challenger and Best Friend. I don't think I even need to explain this to you. Let's dig for fun. Say your Boss is your Best Friend in your work story. Your co-workers or a particularly prickly co-worker may be your Challenger. Jealousy, bullying and gossiping are such revealing traits of Challengers. Remember that fortitude bit? This is an excellent source for it.

3. **Core Group of Friends** are with you for life (even the newly found ones will often last the length). They vacillate between Best Friend and Mentor. It's extraordinary to me to call some of my besties, mentors. Makes me feel evolved. Many of the mentors from my youth, coaches, bosses, teachers are no longer in my life. They played their role (read: discouraged me to make me better/clearer) and left, usually because I was too proud to admit how hurt I was by their "feedback".

4. **Exes/Flirty Friends/Crushes** are my faves because they are unsuspectingly vital to us. They ride the wave between Love and Mentor. This is a special group of people in your life that you feel extraordinarily free to chat to about the most

The Science of Story

personal of details. Hair stylists/colorists. Bartenders. Massage Therapist. Mechanic. Barista. An ex-love. Someone you wish would pick you to kiss so you spill to them on the regular, all with hope of more. We learn a shizzle ton about ourselves from this crew!

I hope you are feeling clarity rumble through your every cell. Maybe a wave of excitement for how your life is your story? Let's get personal and do some exercises just for you.

Style Exercise

Before we get to others, let's reallllly get to know YOU! You are the Star of your Story and these questions will help you see all the wonderful of you. The deeper we connect to ourselves, the deeper we connect to others. The deeper we love and appreciate ourselves, they deeper we love and appreciate others. It's not selfish. It's inevitable.

Character Style Questions	Character: ME
Age and how you feel about being that age	
Gender(s) or Non-Binary or Them/They	
How do you eat? What do you eat and drink? Do you have regular bowel movements? Does a little pee squirt out when you laughs? What makes you sick? What energizes you? Guts you? Ignites? Infuriates?	

Character Style Questions	Character: ME
Do you smell funny? Describe. Good? Bad? Odd? Like smoke? Roses? Vomit? Aftershave? Lavender? Vinegar? Dogs? Babies?	
Style of Talking. Chatterbox or caveman? Southern accent? Brit? Mumbler? Loud? Quiet? Sing-song-y? Monotone? Switches between Yiddish and French? Note sound, diction, quirks, intonations, pauses, no pauses, sayings, and ways of putting phrases together. This is important. This is how the world hears you/your energy/power (or lack thereof)	

Character Style Questions	Character: ME
Style of Dress, including jewelry, glasses, shoes, favorite articles of clothing/accessory? Maybe you always wear the same thing, like a uniform or because you are colorblind, or you are too depressed to care? Or so focused you can't bothered with clothes? Nudist perhaps? What is the one piece of clothing or jewelry that you love? Do you ever wear it? In public? What do you wear to "dress up"? Do you ever? How do your clothes feel to you?	
Style of Movement. How do you walk? Do you use your hands to talk? Do you make eye contact? Are you suffering chronic pain? Do you have "a tell" when you are lying? Ticks? Twitches? Vibe? Eyes closed or open for kissing?	

Character Style Questions	Character: ME
Style of Interaction. How do you relate to others? What do you show love to others? Hate? Anger? Interest? Do you have a persona with others that is different from who you are alone or with close friends? How about with certain family members? What kind of people push your buttons? Are you a toucher or a cringer?	
What is your occupation? What do you wish your occupation was? Both matter, especially if they are not the same. What would you do if you weren't afraid? Does that serve your story? How?	

The Electricity of it All: WHO

Character Style Questions	Character: ME
What are three things you do on a perfect day? Who with? Alive or departed? Animals? Where would it be? What is involved? What does the world look like that day?	
What are your pet peeves? What really bugs you? Willing to share that with other readily? Ashamed? Why?	
What is your predictable reaction to stress? How do you relax? What do you do when alone? In a crowd? On an airplane? In traffic? To celebrate?	

Character Style Questions	Character: ME
What is hard to explain about you? What surprises you about you? What strange, bizarre quirk do you have that doesn't seem to fit but somehow does? Do you avoid certain things about yourself? Which ones?	
What makes people relate to you?	
What sets you apart? What makes us detest or adore you? What is the cost of that trait(s) to you?	

Character Style Questions	Character: ME
Biggest betrayal? Biggest help ever received? Biggest loss? Biggest win? Biggest blunder? Biggest save?	
What do you want to learn? Will you? If not, why?	
What are you willing to risk? What are you not willing to risk?	

Cast of Characters Exercise

Please pick a Story and start laying out your Supporting Cast. I hope you do this for a Story of your dreams. What do you really, really want? Who are the players that will get you there? If that is a wee uncomfortable, do you work life or home life or health life. xoxo

The Electricity of it All: WHO

The Science of Story

Love
Spouse · Parents · Siblings
_____
_____
_____

Challenger

Ex's · "Flirt" Friends · Crushes
_____
_____
_____

Main Character

Bosses · Co-Workers · Teammates
_____
_____
_____

Mentor

Core Group of Friends
_____
_____
_____

Best Friend

The Electricity of it All: WHO

The Science of Story

Love

Spouse · Parents · Siblings

_____
_____
_____

Challenger

Ex's · "Flirt" Friends · Crushes

_____
_____
_____

Main Character

Bosses · Co-Workers · Teammates

_____
_____
_____

Mentor

Core Group of Friends

_____
_____
_____

Best Friend

The Electricity of it All: WHO

Vibration = Voice

I want to take you back to the prior chapter for one moment and remind you that we Express our Style via our Vibration, otherwise known as Voice. Over time and with a lot of practice, I have discovered that for me, it is next to impossible to differentiate Vibration and Voice, much as I may want to at times. Our vibration is our everything out into the world, and our everything within. It is our emotional state of being, which I believe to be our Voice.

Our Voice is how we express *how we feel* about absolutely every single thing, whether we utter a single word, or not.

Vibration = Voice

Voice is...

- the way you speak, sing, moan, groan, make any noise from your vocal cords, including pitch, tone, cadence and volume
- the way you use your body to move, speak, exercise, share intimacy, flee, flight, rest, be
- the way you tilt your head, raise your eyebrows, roll your eyes, and so on. All those little micro-moments you do
- the way you listen, hear and react to everything
- the way you write everything, the words you use and the order you put them in and what you decide to write about
- the way you repeat, mock, gossip and share
- the way you divulge, announce, proclaim and vent
- the way you articulate and enunciate
- the way you speak without using your words
- the ways you remain silent
- the choices you make
- the opinions you have and the ones you share
- the attitude you carry
- your view of everything
- your instrument for who you are and what you believe in
- the mouthpiece of your soul

Starting to feel it? The. Power. Of. Voice. Is. Extraordinary. It vibrates all around you. And is you.

Hone your voice, hone your life. In perpetuity.
It is a matter of practice.

Vibration = Voice

Honing Your Voice Exercise

Please sit quietly or have a walk in the woods or the beach and start to inventory moments from your past. You don't need to write them down unless you want to. They usually don't leave once you've picked the scab.

Let me see if I can help stir things up…

- Death of a Loved One or Beloved Pet
- Getting engaged
- Assault/Violence done to you or by you
- Divorce
- Breaking up
- Abortion
- Adultery/Cheating
- Moving away
- Miscarriage
- Fight where you were completely abhorrent
- Loss of a home
- Birthday
- Fight where someone was completely abhorrent toward you
- Quitting or being fired from a job
- Anniversary
- Car accident
- Fire
- An extraordinary event
- Sporting accident
- Burglary you were victim to or committed
- Seeing something you created purchased and/or published
- Interaction with the police or other authority where the stakes were high (school principal, judge, military,

- TSA)
- Sexual assault to you or by you
- Birth of a child
- A big award you won
- Getting married
- A contract you signed

Got a couple? Great! I would now like you to **write a letter** to the person(s) involved in the situation.

I encourage to write the letter from the you of then. If the event happened when you were twelve, be that twelve-year-old.

Write as many letters as you like. To me, this is you **reclaiming your voice.** There is always more healing that can happen. Always more to learn, to become, to explore. Please write it as you *felt* it. Whatever you need to say, however you need to say it, go for it. Be brave and face this. Capture it to release it. We dance around these moments. I'm asking you to dance with them.

Extra credit

Write the lesson you learned from the event. Tell yourself how you feel now at your current age. Dare I suggest, celebrate it?

I recognize this is difficult. I have done it. It took me to my knees. And remains the best thing I ever did for myself, and those I love.

You are a superhero! WELL DONE!

Vibration = Voice

Letter to:

The Science of Story

Lesson/Message Received/Resulting Transformation/Triumph:

You are seen. You are heard. You are loved.

Vibration = Voice

Letter to:

The Science of Story

Lesson/Message Received/Resulting Transformation/Triumph:

You are extraordinarily brave.

Vibration = Voice

Letter to:

The Science of Story

Lesson/Message Received/Resulting Transformation/Triumph:

You are right on time. Always.

Vibration = Voice

Letter to:

The Science of Story

Lesson/Message Received/Resulting Transformation/Triumph:

You matter.

Vibration = Voice

Letter to:

The Science of Story

Lesson/Message Received/Resulting Transformation/Triumph:

You are FREE. You are BEAUTIFUL.

Vibration = Voice

Letter to:

The Science of Story

Lesson/Message Received/Resulting Transformation/Triumph:

You are LOVE.

The Principle of Reason: WHY

This brilliant man, Gottfried Wilhelm Leibniz, reminded us that everything has (must have) a reason, or cause.

Why bother if not for a reason? Trees are here for photosynthesis. Oceans for weather, cooling, managing our flow. Obstacles for solutions. Sex for perpetuation. And so on and go forth.

The Reason WHY is the energy pulsing rhythmically within us and our stories. You are the potential energy, WHY activates it.

In other words, no Why, no Story. Why do we want to be happy? Because we are sick of feeling sad or afraid or both. Ideal=Happy. Breakthrough/Opposite= Sad/Fear-filled.

Why INSPIRES.

The Science of Story

The Why captains the ship. And if you don't know WHY then you may be a little adrift.

When we share our stories with each other, we often leave out the why. Either we don't know the why or we are afraid to verbalize it. We talk about HOW we did something or WHAT and WHERE we did something. But we veer wayyyy around WHY we did something. Or want to do something.

The Principle of Reason: WHY

It's such a shame because while the HOW, WHAT and WHERE matter, the WHY is the juicy, driving force. It is the underlying, inescapable force for e v e r y t h i n g.

Why=What Your Story is About=
The Journey from Breakthrough to Ideal Traits

What is it about your specific choices that compels you to the core of you? That's a Why. That's the force of nature pulsing with you. It is natural for you to be driven by more than circumstance and tangibles.

WHY is what YOU need to discover at all costs.

Remember those obsessions/passions? Why are they obsessions? The more you share THAT why the more you will draw your dreams, clients, buyers, bosses, lovers, fans straight to you. Why? Because they share your passion. The freaking can't help it! They HAVE TO know more because **their why matches your why.**

Tell Me Why Exercise

1. Grab a few or all of the obsessions you captured prior to this. If you didn't do that, could you please write down at least 5 things YOU ARE VERY PASSIONATE about.
2. For EACH passion/obsession, write WHY you are obsessed/passionate about it. Why does it matter to you? Why do you care? What is it about YOU that makes THAT so important? And why is that?
3. Can you capture some ways that you would share your why? Maybe a little video? An article? A book? Hmm. Just capture the idea for now. (Unless you are so inspired you have to get on the sharing NOW! I love it! Go for it.)

We created a table format you may want to you to use to capture yours.

The Principle of Reason: WHY

Passion/Obsession	
Why do you care about this passion?	
What is it about you that makes it important?	
Why?	

Passion/Obsession	
Why do you care about this passion?	
What is it about you that makes it important?	
Why?	

The Principle of Reason: WHY

Passion/Obsession	
Why do you care about this passion?	
What is it about you that makes it important?	
Why?	

Passion/Obsession	
Why do you care about this passion?	
What is it about you that makes it important?	
Why?	

The Principle of Reason: WHY

Passion/Obsession	
Why do you care about this passion?	
What is it about you that makes it important?	
Why?	

Passion/Obsession	
Why do you care about this passion?	
What is it about you that makes it important?	
Why?	

Passion/Obsession	
Why do you care about this passion?	
What is it about you that makes it important?	
Why?	

Passion/Obsession	
Why do you care about this passion?	
What is it about you that makes it important?	
Why?	

The Principle of Reason: WHY

Passion/Obsession	
Why do you care about this passion?	
What is it about you that makes it important?	
Why?	

Passion/Obsession	
Why do you care about this passion?	
What is it about you that makes it important?	
Why?	

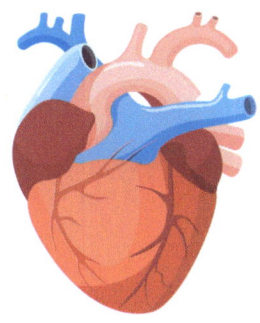

Let's Postulate: THEME

Hands down my favorite scientific activity is to POSTULATE. I love it, love it, LOVE IT! It's the playful side of science. It's when we advocate the existence of something based on reason, discussion or belief. A postulate is a guide for which we can base heaps of other things. For instance, Einstein's postulate of special relativity suggests that any law of nature should remain the same at all times, regardless of who is measuring them. I agree.

A Postulate=a Principle=a Theme.

It's what you rely on, and apply to yourself (and your stories) without question, giving you the freedom to become who you are more and more.

Theme is likely a term you've heard of. But maybe haven't spent a lot of time thinking about?

Let's think of theme in the way it is most often used: as a song for a movie. For instance, the Jaws theme song. Dah Dun. I just shivered. Absolutely daunting. The message is clear: let's not mess with great white sharks. They bite. And win. Let's let great white's be great white's. Two notes remind us there is plenty of ocean for us all. How about Rocky and The Eye of the Tiger? Rocky's Why is to prove to himself that he is a world champion boxer. That he CAN fight and WIN. Obviously, he HAS to have The Eye of the Tiger.

The THEMES of your life = The PRINCIPLES of your life.

Theme represents our belief systems. Theme holds WHYs at 10,000 feet. Therefore, theme is always there, surrounding, needling, biting, and chirping away at your story/you. Often we are completely oblivious to our themes. But they are always there.

Theme is what we live for. And what we sabotage ourselves with.

Understanding your themes, then leveraging them, will make creating/living more fulfilling and free flowing.

Theme is your best friend.
The one you constantly check in with to be sure you are on track.

Let's Postulate: THEME

- We MUST BECOME our TRUE SELF.
- We MUST believe in POSSIBILITY.
- We MUST treat each other with RESPECT.
- We must PROTECT our CHILDREN.
- We MUST FIGHT INJUSTICES.
- We must NEVER CONFORM to AUTHORITY.
- We must FORGIVE OURSELVES.
- We must LOVE OURSELVES.
- We must FIND OUR VOICE and USE IT. *(This is one you have, trust me.)*

Themes are things you believe so fully, you know with all that you are that you CANNOT, and WILL NOT COMPROMISE on it ever, never, ever.

Here comes the best part!

Drum roll...

We CONSTANTLY compromise our principles/themes.

We betray our non-negotiables, and it nearly kills us.

That's called LIVING.

Let me show you how this works and you will be nodding, I promise.

We MUST live our lives to the FULLEST. *(Ideal)*
Therefore, we DO NOT LIVE OUR LIFE to the FULLEST. *(Breakthrough)*

We are scared to leave the house. We are addicted to our fears. We numb ourselves.

We are lonely. We don't take the trips we dream of. We hide. We remain small. Until we don't.

Until something happens that forces us to change: We get cancer. We lose the house. Our spouse cheats on us. We get fired.

Here's another:

We MUST be AUTHENTIC.
Therefore, we LIVE in LIES.

We block anyone, everyone and everything from knowing/seeing the real us. Pretender. Fraud. Lost. Imposter.

We hide in the wrong gender. We starve ourselves. We stay at our corporate job and never open that bakery. We stay married to an abusive spouse. We acquiesce, we compromise, we give up on ourselves.

Until we don't.

We meet someone who had their top surgery and feel inspired to make an appointment. We walk out of the house and drive to a crack house to get high for the first time. We get laid off from our corporate job and decide to bake a cake. We shift. We heal. We thrive because we live as our truest selves.

We MUST treat each other with RESPECT.
Therefore, we must treat others like CRAP. And suffer for it.

Let's Postulate: THEME

We are a bully. We beat our spouse. We hate. Until we don't. Our son gets beaten up by a bully. We lose our fortune. We get caught plagiarizing our memoir. We run a Ponzi scheme and get caught.

We must PROTECT our CHILDREN.
Therefore, we must harm children, or have been harmed as a child.

You tell me now... go ahead, don't be shy. Write it down if this is a theme that fits you.

We MUST FIGHT INJUSTICE.
Therefore, we must have been unjust or been deeply scarred by injustice.

We MUST believe in POSSIBILITY.
Therefore, we must LIVE BY THE RULES and NEVER do ANYTHING that has RISK.

For instance, we are living with our parents for our 45th birthday. We have never driven a car. We have never earned our own money. We don't go anywhere or do anything alone. We write books but never share them, much less try to publish them. We dream but never do. We don't even dream anymore. We live through and for our children/husband/friends. We stopped baking. This is such a good one!

We must NEVER CONFORM to AUTHORITY.
Therefore we must CONFORM often and about really important things in our life. Job. Home. Family. Friends. Religion. Social rank.

The Science of Story

We must FORGIVE OURSELVES.
Therefore, we must not forgive ourselves, or others. We must not be able to forgive. And be as rigid as possible about it. Tension creates resonance.
*Always a good theme in family stories.

We must LOVE OURSELVES.
Therefore, we must hate ourselves, and take it out on others.

We must FIND OUR VOICE.
Therefore, we must BE UNABLE TO EXPRESS OURSELVES.

Theme, like its BFF Why, MOTIVATES, INSPIRES and PUSHES you/your clients/characters to do things they might not otherwise do—good and bad! It is what makes your story BELIEVABLE and RELATABLE.

Let's Postulate: THEME

THEME SONG Exercise

As promised, let's have it. What's your theme song? Yes, you can have more than one. Note for yourself what each one represents. Before I knew myself and had honed my voice: *Broken Wings* by Mr. Mister. After, as in now, *Sailing* by Christopher Cross. My novel series' main character, Truitt Skye, has the theme song, *Wondering Where the Lions Are* by Bruce Cochran. The theme song of this book is *Peace Train* by Cat Stevens. I never loved these songs. They were not my favorite songs. I didn't know the words until each came to me without hesitation and wouldn't leave until I looked up the lyrics and sat in uncanny awe of the connection. That's why I believe in the power of a theme song.

NON-NEGITIABLES Exercise

Please grab your pen and paper or computer and list out **your non-negotiables**. Give yourself 15-20 minutes to jot them down and then pluck out one or two that you are super-charged by and expand on why they are your non-negotiables. What happened in your life experiences that makes them MUSTs and MUST NOTs. Take your time. Go deep. Be brave. You don't have to share this with anyone, but I sure hope you do. And if you want to take it even further, what stories or programs or teachings or sharings are coming to mind as you write your non-negotiables?

Remember to use the prior work you have done to help you. Grab those IDEAL TRAITS. And BREAKTHROUGH traits. They are your starting points.

If you are having trouble, write down the biggest betrayals **or traumas** of your life. The BREAKTHROUGHs live there. You can also look at your greatest triumphs and find the theme.

For instance, I was involved with a man who was seeing other women. HONESTY is one of my ideal traits. Therefore I must be lied to, I must lie to myself and to others, which I did in this relationship. I knew something was wrong. I knew he was seeing other women, but I never voiced it. I hid because I was afraid to lose everything, which I did. Kinda. This leads to another of my Ideals: FREEDOM. Therefore I must be IN CHAINS (Breakthrough). And so on and go forth.

I know it is hard work. It is worth it. You are worth knowing yourself this deeply. The other side is MAGIC. I promise.

Let's Postulate: THEME

Theme 1

IDEAL:

BREAKTHROUGH:

THEME=

Theme 2

IDEAL:

BREAKTHROUGH:

THEME=

Theme 3

IDEAL:

BREAKTHROUGH:

THEME=

The Science of Story

Theme 4

IDEAL:

BREAKTHROUGH:

THEME=

Theme 5

IDEAL:

BREAKTHROUGH:

THEME=

Theme 6

IDEAL:

BREAKTHROUGH:

THEME=

Theme 7

IDEAL:

BREAKTHROUGH:

THEME=

Theme 8

IDEAL:

BREAKTHROUGH:

THEME=

Theme 9

IDEAL:

BREAKTHROUGH:

THEME=

Theme 10

IDEAL:

BREAKTHROUGH:

THEME=

Let's Classify: GENRE

Classification is grouping like things, yes? Domain, kingdom, phylum, class, order, family, and finally species are the taxonomic levels for plants and animals. Us humans are Animalia, Chordata, Mammalia, Primate, Hominidae, Homo, sapiens, Human! Story breaks down this way as well. The prior data—chapters—led us straight here and in the next chapter, we will dissect Story. This chart goes from broad to narrow.

Domain → Legends
Kingdom → Beliefs
Phylum → Emotions
Class → Thoughts
Order → Experience
Family → Genres
Species → Stories

The Science of Story

Let's chat about Legends. These can be memories that belong to you from childhood or adulthood. They are overarching powerhouses that we accept, rely on, blame, avoid, hide behind and often might be best left in the past unless they support you in every sense of the world. Think cultural rules, nationality traits, religion, gender. They are not in any way tangible, yet they often control us. Like I said, powerhouses. Can you think of some of yours? Oh yah. I bet you can. To me, the biggest legends aren't dead. They arrived in our childhood. Some of us don't *think* they have a hold on us, but they do. Others of us think we can't get past them, but we can.

I'll give you an example or two, illness and/or death of a sibling at a young age (anything under say 60, but even more powerful if the sibling was a child and the fall out and hold it had/has on you and your childhood memories). Feeling responsible for the harm or death of another, as in suicide, an accident, abortion, miscarriage or disappearance. Divorce is a biggie. No matter the details, if it happened, it is one of our legends. And fuels our beliefs about what marriage is, about men, husbands, wives, women, honesty, integrity, self-worth, freedom, commitment and possibly more. By the way, having parents that are married for decades is also a legend where nearly the same beliefs formed. Interesting, hey? Ponder it and see what you believe.

Legends run so deep within us we can't understand their reach at times. They are the ghosts that haunt us until we transform them into appreciated experiences that served us well. In order to do that, we have to poke around the next layer supporting our stories: Beliefs.

When we believe something, it holds the highest value to us. It is nearly impossible to buck someone's beliefs. They are the platform, as in far bigger than a meager layer, for our stories. It's imperative to be mindful of what we believe, remembering that

100

they are first formed as a response to our Legends. Once our legends have become our friends, our beliefs start to wriggle free to possibility.

I'm not suggesting you talk about your ghost stories every day. Quite the opposite. However, we can't pretend they aren't there or lie about them if we want to live the stories we want to live. Story is about transformation from our yesterdays. We need to find a way to include them, so they no longer have power (read: distracting pain and angst) over us. Transformation relies/demands on us learning from, and letting go of, our Legends. Only then can we figure out what we BELIEVE about our now. In other words, yesterday is long gone, unless you perpetuate it. The moment you go digging around in the worms and debris of your yesterdays, that is all you can see.

Can you imagine a million bucks? Can you see it? What does it look like? A check made out to you is lovely. In pennies? Excellent. Thatsalottapennies. Bills in the palms of your hands? Incredible! Whatever pops to mind is right. Can you feel it? Or are you filled with aloofness or hate toward money? Or are you in love, deep, rich intimate love with all that moolah? Do you believe you can have it? A million? Two? Ten? How about a book deal? True love? A home you feel safe and happy in? A good friend who asks you how you are for no reason at all? A decent drink of clean water? A birthday candle to blow out? The ability to read and write? To live in the body you desire? To go to work excited and fresh every day? What do you believe? It's so darn important because:

> What we believe we are, we are.
> What we believe will happen, will happen.
> What we believe matters, matters.

Onto the next classification, the tertiary level, Emotions. The ways we believe or don't believe in ourselves and our dreams and desires, dictate our emotions. It's hard to believe that our feelings have two thick layers of us to get through when they are so often right at the surface! I believe that is because our Legends and Beliefs are quite tightly ingrained, like the cement in the foundation of a home. Feels like big stuff, doesn't it? It's just Story. Knowing how it works can make all the difference in your world. xoxo

When we believe that death is the worst thing that can ever happen, we feel devastation, sadness, horror when we hear about death. When we believe death is natural, we feel at peace, hopeful in the happiness of the departed, appreciative of every moment we have and ever had with the departed. When we believe death should not be discussed, we feel cut off, lost, fragmented, lonely. When we believe that death is unfair, we feel angry, rejected, aggressive, wronged, righteous. When we believe that death is a release, we feel grace, ease, relief. See the rhythm between our Beliefs and our Emotions? See any connection to your Legends?

Our Emotions fuel our Thoughts. When we feel, we think about it. We ask our brains to find words and sentences to express them. What good's a feeling without doing something with it? Sharing. Understanding. Appreciating. Prolonging. We have to do something with our feelings because we want to make them real/heard/seen/known. Social media thrives because of this essential component of human nature. It's an instant way to put our thoughts to our feelings, making it addictive to the very core of our beingness. If a tree falls in the woods, does it make a sound? Not unless someone is there to hear it. Without thought, every possible outcome exists, and the feeling fades almost as fast as it arrived. Once we give it a thought, we made a choice about what the feeling means.

Let's Classify: GENRE

Thoughts give way to Experiences. By this I mean our actions, choices, knee-jerk reactions, inclinations, preferences, and habits. In essence, the life we live is a product of our thoughts. Change your thoughts, change your life. Love your thoughts, love your life. Hate your thoughts, hate your life. And so on. Because of this, what we think, we become. Let that sink in for a sec. Whoa.

Pause for deep breath. xoxo. We finally arrived at our destination: Genres.

Genres are beautiful groupings of our Experiences. It's more than we can handle to process every single experience going on in our lives as individual, unique moments. Though that is what they are, my beautiful friends. We are so eager to speed up, press on, move forward, we forget to pause and regale at each second of our precious life. I'm not judging, just noting for the purposes of understanding Story. We group to help our brains. We actually naturally classify and group all.the.time!

Genres are represented by or depicted by symbols. Yes, you read that right. Symbols. Neat, huh? We're talking about the colors, pictures, music, smells, shapes, and sayings we use to express who we are and what matters to us. If you have hearts and pinks and flowers surrounding you, guess what? Love is one of the genres in your story. If you have family photos hanging, books on relationships and marriage, quotes framed on the walls or throw pillows with things like "Bless this Happy Home" "Family First" or "Love. Laugh. Learn.", some your stories fall into the drama genre. If there are stars on the ceiling of your room, books about science, nature/Earth photos, then Science/Nature is a genre you love. Are you following? Have a look around. What symbols do you find?

The Science of Story

I'll toss out a few of the common genres to get you going, but you can get far more specific than I can. And you will have a set of genres that define the majority of your stories.

Love/Romance	Drama/Family	Career/Work
Science/Nature	Fantasy	Books/Knowledge
Horror	Detective	Pets/Animals
Comedy	Art	Ocean/Beach/Sailing
Religion	Travel/Adventure	Health/Wellness
Crime	Culture	New Age/Tarot

From here, we can drill down a broad category, like Art, to Modern Art. Then perhaps more specifically, Korean Modern Art. This potentially adds the Culture genre. I would be able confirm that if there were other Korean symbols reflected in your space and some K-Pop playing. It gets super interesting as you realize how much of your story is in every part of your life. Symbols are FAR more powerful than words. Like Genres, they lump our layers together perfectly and obviously.

Once we add language to our symbols, we are creating Stories. If you aren't feeling like a Storyologist yet, I'm shocked! We have become quite intrepid scientists of story, have we not? In the next chapter we will get super serious about the Guts of Story. But first, some super fun homework!

DEFINING GENRE Exercise

Please make a list of your LEGENDS. Who or what haunts you?

Why stop with LEGENDS? **Please jot down some of your BELIEFS,** and as you do, please think about (and write about, if you like) how you FEEL about them? If you are having some resistance, draw from your EXPERIENCES. How do you feel about Commitment/Marriage? Food/Health? Behavior? Politics? Authority? Parenting? And so on...

Let's Classify: GENRE

SYMBOLS List

The Science of Story

And last but never least... What are the GENRES of Your Life/Stories?

Let's Classify: GENRE

Extra Credit: CLASSIFICATION Exercise

Match up your information into the table for one or two of your Genres. I'll do an example for you to get you started

Domain → Legends	Divorce of parents at a very young age (under 3)
Kingdom → Beliefs	Relationships are hard. Relationships don't last. People cheat and are cruel to each other. Marriage stops freedom. Heartbreak makes you bitter. Leave them before they leave you. Men are not to be trusted. Women ruin men.
Phylum → Emotions	Confusion. Needy for love. Needy for attention. Anger. Sadness. Loneliness
Class → Thoughts	I am not good enough for love or marriage. I'm damaged/have baggage because of my experiences. I don't deserve to be happy like couples who make it look easy. I can't be loyal to only one person. That never happens. Men don't love me. Only needy women are attracted to me.

Order → Experiences	Break Ups Cheating Self-Sabotage
Family → Genre	DRAMA
Species → Stories	This child was too trusting because she wanted to be loved and seen and held. Her parents were too caught up in their own junk to pay attention so this child pays extra close attention to how people feel—more than caring how she/he feels, they want others to be okay. This child is angry for years and years and short-tempered, but it is a learned anger, not theirs. The deep sadness this child has is also not theirs. It is learned and the greatest triumph of this child's life. They know they are an adult when they take responsibility of their life and let go of the anger and the hurt and the loneliness and the sadness that was never theirs and fall into deep, magical lasting love with the partner of their dreams. They may have a marriage or two fail along the way, but this is the transformation on offer—the true triumph to this story, which creates A NEW LEGEND for this child's children.

Let's Classify: GENRE

Domain → Legends	
Kingdom → Beliefs	
Phylum → Emotions	
Class → Thoughts	

Order → Experiences	
Family → Genre	
Species → Stories	

    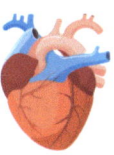 

The Guts of Story:
WHAT, WHO, WHEN, WHERE and HOW

In this section we'll be talking about the nuts and bolts that construct story. This is nothing new. Great scientists mapped this for us long ago. The elements of story have been around as long as humans have been communicating. I have my own little twist on them, but you can thank Aristotle for capturing the same elements of storytelling taught today in 335 B.C. Joseph Campbell's, *The Hero's Journey,* remains among of the best for men. Maureen Murdock's, *The Heroine's Journey* is a must-have for women.

I like to keep things as simple as possible so that everyone can incorporate them into daily life and the lives of your beloved family, friends, clients and colleagues. Ideally, these story points will become second nature, called upon in your everyday observations, and most important, guide you fully in your life.

Indulge me please for a moment to define what I mean by story. First and always, a story is contained to a brief beginning, a messy, long, confusing, hard middle and a quick, often forgettable, end that leaves the door open for growth and expansion.

Let me reiterate, the beginning and endings are BRIEF, as in, not long. The heart of your story lies in the muck that comes in between the start and finish.

Act One: Birth.
Act Two: Life.
Act Three: Death and/or ReBirth, depending on your beliefs.

The beginning is but a glimpse of what could be. It's nothing more than a question, the infamous one being, "What if?" Good endings, the ones we long for, root for, ache for, only last a few seconds. He said yes! She said take a hike! I drove the new car off the lot! Seabiscuit won! I'm engaged! We're alive!

If you think about it for a sec, there is next to no difference between the beginning and the ending. It's a cycle. The bulk of which lies in the middle. As soon as we get what we want, a new idea springs forth. In fact, thousands of ideas occur daily, as do triumphs. (You woke up, you went to the bathroom, the sun rose, you ate food. Must I go on? You are doing so well all day long, you hardly notice. Triumphant endlessly! Well done!)

The following graphic is offered to help you see and hopefully retain the classic steps of Story, which I lovingly named, The Guts of Story. It's a bit of a play. On words. These steps are, in fact, the innards of every story but more important, it takes GUTS to live your story. xo

These are general steps. As I teach on a topic, for instance, a Love Story, for fiction,

memoir or film/television, I start with these steps, then expand to 14 steps, then the 21 steps, then I create a full scene-by-scene list which is usually 75-100 individual steps. Further, every chapter/major scene of a novel/script often contains all seven steps as a mini-story within the larger story.

I developed this broad version for SoS (The Science of Story, the book you are holding) with the intent that it could be memorized, and recalled, as needed—especially when someone I love needs to remember that TRIUMPH is always the outcome of a full story. However, I do remind us all that New&Shiny is always preceded by being All In, which requires us to "confess" and "relax," which are pretty much two of the hardest things we ever do. That is until you see the power of letting go of anything and everything that holds you back. Once you get that, you'll start confessing and chilling on the regular with ease and humor and love.

My point is, these steps are important to your life. To your daily life. To you creating a life you love. You matter. Your stories matter. Your stories are matter. They are "the stuff "of you, the materialized energy of you. We ARE the stories we tell. These steps help you define and connect to everything and everyone. Knowing where you are in your story is the secret sauce to your dreams coming true, regularly and repeatedly. What could be better?

The Science of Story

The Guts of Story: WHAT, WHO, WHEN, WHERE and HOW

What Envisioned

1. Possibility (Act One Starts)

This occurs when we daydream. When we see ourselves as we want to be seen. When we let our minds and hearts wonder and possibilities arrive. The ones that really matter (and want to become matter) will not go away. Think about those passions again. A trip you want to take. The husband or wife you want to marry. A garden in your backyard. A novel. A memoir. Eating chocolate cake from a favorite bakery in a city far away from you. A new business venture. A new line in your business. Sharing a recent experience that changed your life with others in hopes of inspiring them to dream more and be more. Him. Her. A puppy. A kitty.

Possibility is the first scent of a story. Sometimes it hits like a bolt of lightning and leaves a mark you can't ignore. Other times, it comes in like a faint whisper for years on end.

Possibility abounds. And ALWAYS begins a story.

The Science of Story

Who/Theme

2. Today (Act One Ends)

Then we reel, yank, and push ourselves back to reality, and reality does not have room for such whimsy. You believe your dream is something that **happens** to someone else. Daily life creates immediate friction in your life.

Doe-Dee-Doe is the life blood of the human experience. Wake up, do morning routine, go to work or school or back to bed, eat along the way, sleep, and repeat. Something more exists. You know it. You just daydreamed it! But it's **not** for you. How could it be? This is GOOD NEWS! The whole point of a new story is to challenge our Today.

The Guts of Story: WHAT, WHO, WHEN, WHERE and HOW

When and Where

3. The Sign (Act Two Starts)

This is what you've been waiting for, whether you know or want to admit it or not. Something happens in your life and you can't go back. It's powerful, not subtle.

We get fired. We get a text or email or phone call that rocks us to our core. We catch our spouse cheating. We win the lottery. We get a book deal. Our mother dies. We lose the baby. We come out to our family. We get married. We adopt a baby. We buy land to build a house. These are BIG events folks, not small. We inherit a chunk of money. We're found guilty at trial or accused of murder, depending on your why.

The Sign is NOT when we meet a stranger on a train, unless that stranger starts shooting everyone or confesses that he is your father, or both. Harsh, I know, but I'm trying to make a point that gets missed a lot. It is not buying a URL for your new business. It is not writing the prologue to your novel. It is not signing up for an online dating service or business course. It is not having a photoshoot or going to an audition. It is not saying yes to a date, proposal, job offer. Those are middle moments.

The Sign is not a task in your plan, my loves.
It HAS to be a MASSIVE moment because of what happens next.

The Science of Story

Why/Theme

4. Resist to Persist (Act Two)
Welcome to the MIDDLE of the story. You say: No way! No how!

You do NOT feel capable that you can do this.

This is a period of one step forward, ten steps back. You get a little traction but not nearly enough to commit to the new you. You find rhythm then it is gone. You can't seem to quite get what you want. Maybe you can't get started. You spin. It's mental. It's rough. It's scary. You try and fail. You are exhausted. No one is supporting you the way you need it. You work so freaking hard and can't catch a break.

Not too worry, loves, because in the next step. . .

The Guts of Story: WHAT, WHO, WHEN, WHERE and HOW

How

5. Help Arrives (Act Two)

This is usually in the form of a someone. A mentor. A life coach. An acting or athletic or dating coach. An agent. A real estate agent. A new friend. A parent. A teacher. It can be a course, a school, a community, but likely there is a standout someone. This is Yoda. And it is Darth Vader.

That which antagonizes us, prods us, or if we're lucky, propels us forward. That is why I mention the "bad guys" as well as the good. Help is often in that nasty a-hole at work that keeps stealing your ideas. It is your mother. It is your father. It's that bully you need to stand up to. And it is your friend that never leaves your side.

**You are the Help that arrives into a person's life.
And you need Help to Arrive in order to continue to grow, as well.**

It is POWERFUL in story and in life. Asking for Help is a superpower. One that is dwindling in humanity. I am guilty of thinking I can do it all myself when I absolutely cannot, need not and want not. Let them say no. Let them know it is okay to say no. BE OKAY if they say no. There are billions of people on this beloved ball of ours.

There are many, many folks who can help. May you not limit yourself to who is familiar or paths that you know. Ask for people who possess the expertise you seek, be willing to pay for advice, be willing to charge for it once you possess it, and be willing to share the credits as you celebrate making humanity, our fellow brothers and sisters who are so much more like us than not, a better species.

Take responsibility and do not overstep bounds, beg, or undervalue the mentor who has arrived to propel you forward to the change(s) you desire. Especially when that mentor is you.

There are many, many folks who can help. May you not limit yourself to who is familiar or paths that you know. Ask for people who possess the expertise you seek, be willing to pay for advice, be willing to charge for it once you possess it, and be willing to share the credits as you celebrate making humanity, our fellow brothers and sisters who are so much more like us than not, a better species.

Take responsibility and do not overstep bounds, beg, or undervalue the mentor who has arrived to propel you forward to the change(s) you desire. Especially when that mentor is you.

The Guts of Story: WHAT, WHO, WHEN, WHERE and HOW

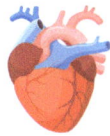

How/Theme

6. All In (Act Two Ends)

Here's what happens when you feel supported: you COMMIT to becoming more than you were when the next thing in your life was but a Possibility. Story means transformation at its core. It's not a story if nothing changes.

Once you are All In, you are no longer able to quit.

The thing to remember is that before we commit and are All In, we will face one more obstacle and it is BIG. You will HAVE TO make a BIG CONFESSION/HAVE A MELTDOWN and QUIT/WANT TO QUIT, then you realize it's too late. Can't quit. HAVE TO PROCEED. It always involves a big secret being revealed or preferably confessed and RESOLVED. You let go. You relax and ACCEPT YOURSELF. When we are ALL IN, no secret matters anymore. No matter what, we are sticking with it until you are New and Shiny and your Dream has Come True.

The Science of Story

What Actualized

7. New and Shiny

There are days, weeks, years of being All In before the New and Shiny truly takes hold. In order to be New and Shiny, the change must become habit and fully integrated into you. That said, we usually only experience a glimpse of this. We don't dwell here. We're off on a new story before we celebrate most triumphs. Years, or at least months, of trial and tribulation lead to about a minute of New and Shiny.

We're going to move on now. It's a lot to digest. Please give yourself time to let things sink in. It doesn't matter if you get this wrong. You have the power to fix/change anything.

The Guts of Story: WHAT, WHO, WHEN, WHERE and HOW

Guts of Story Exercise

Please take a few moments to write the seven steps of an event in your life. A good one is the story of your first kiss or of your engagement. Or perhaps when you were betrayed.

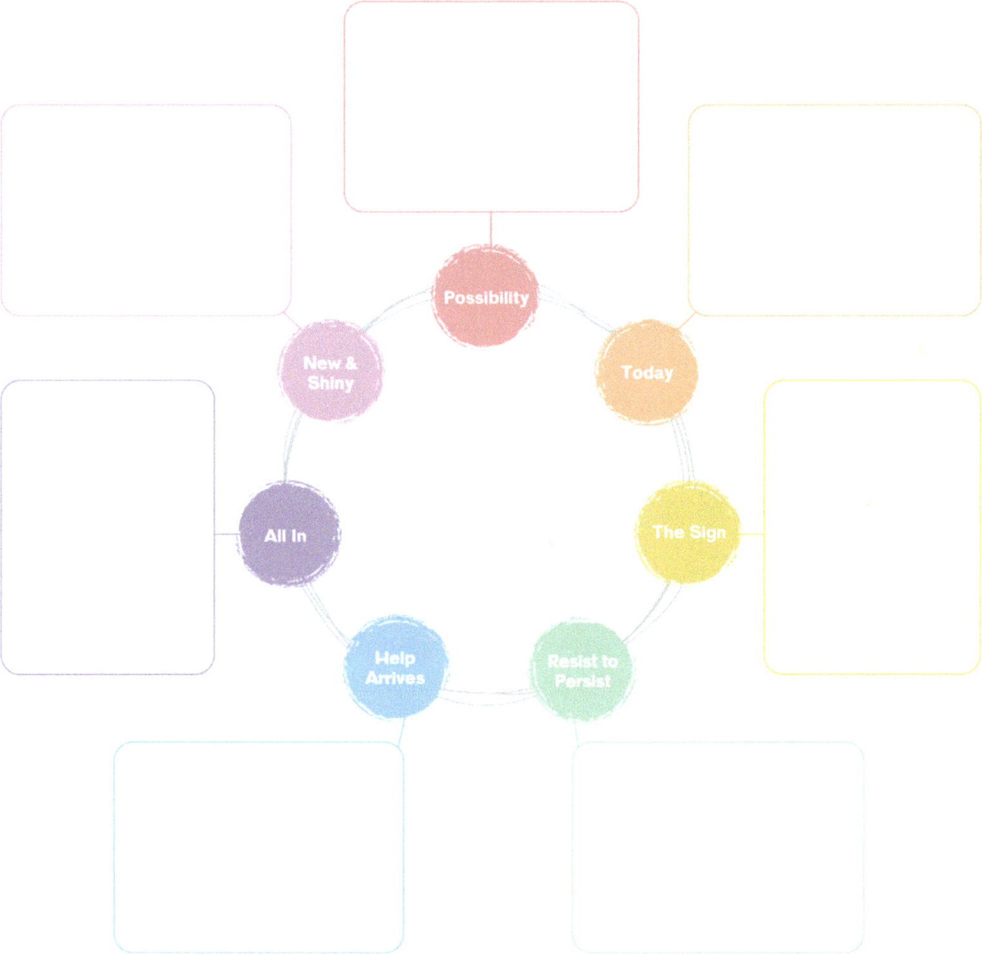

The Science of Story

- Possibility
- Today
- The Sign
- Resist to Persist
- Help Arrives
- All In
- New & Shiny

The Guts of Story: WHAT, WHO, WHEN, WHERE and HOW

Extra Credit

What is your favorite film or book? Can you re-watch/re-read it and note the seven steps of story here? Our favorite films/books are mirrors to our stories. What magical insight it is to recognize yourself as the hero/heroine of your life/story through the imagination of another. Resonance is power.

The Guts of Story: WHAT, WHO, WHEN, WHERE and HOW

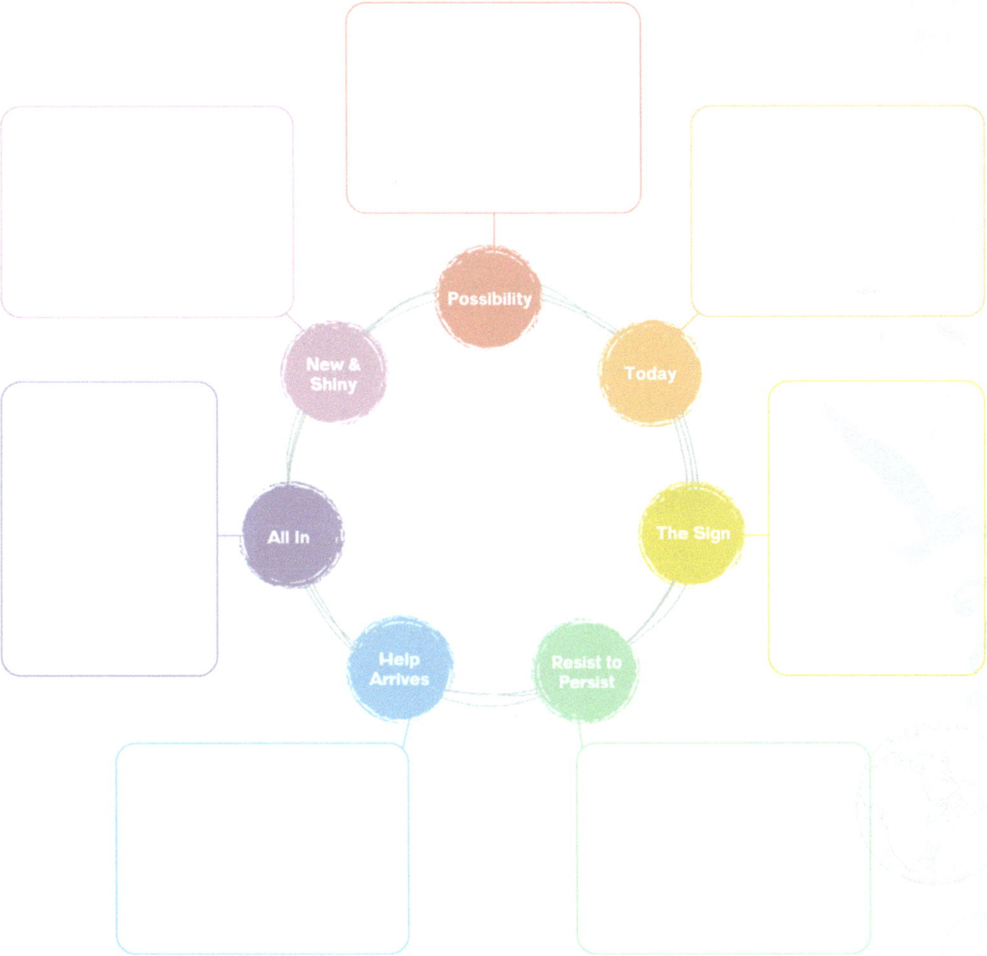

- Possibility
- Today
- The Sign
- Resist to Persist
- Help Arrives
- All In
- New & Shiny

The Science of Story

Transition Time: THE END

I hope you are still with me and feeling incredibly eager to get busy at making your dreams come to life. I believe in you, completely and utterly. By the time you confess and reveal your big secret, the transition to the new you is a foregone conclusion. There is no longer a need, or possibility, for you to turn back. The only way is forward. All you had to do was spit it out. Well done!

We must release our hidden hauntings to step forward on our path. We are infinitely layered. Always more to let go of. Always becoming more. I have found in my own life that I have big dreams (being a published author) that I had to admit to wanting several times before the deal(s) arrived. I'm much, much better at confessing now because I know how awesome life is once I do. Be better than me.

Once you are All In, massive growth, massive success, massive new energy and birth/rebirth are fully on offer. Act Three, the End, is where this newness takes form, and then of course, a brand-new story/desire/dream begins immediately, creating a beautiful, cyclical rhythm to life.

> Always a dream, always rejection of it, always help arrives,
> always triumph, always transition, always a new dream begins.
> That. Is. Story.

Bringing it Together

Let me share an example to give you the full picture. CONNECTION is a key ideal trait for me. And my breakthrough trait to it is SEPARATION. These are life themes for me. Therefore, all the stories of my life—big and small—touch on this theme. Sometimes prominently, sometimes subtly. But they never go away. They are me.
Ultimately, I want Connection. Deep, meaningful, joyous connection to my friends, family, mate, work, land, home. . . Everything! And on bad days/decades, I need to **breakthrough** the feeling that I am Separate from those. I am someone who will not talk to a loved one for months or years. I am someone who will be nomadic. I am someone who will be left unexpectedly. I am someone who will be alone, a lot. I am someone who will separate with ease and leave what and who I don't connect with behind without guilt or regret. I am someone who can be very detached, and very attached. I am someone who must be abandoned many times.

Why? Why are these things true of me?

So that I can REALLY KNOW and EXPERIENCE CONNECTION. If I didn't value Connection so much, I wouldn't experience abandonment. It simply would not happen in my life. There would be no point.

Transition Time: THE END

Ideally, I am someone who will be so available and open to all that I will fill myself and others with unending, deep, nourishing, gorgeous love. I am someone who will have extraordinary conversations with others. I am someone who will love and honor the Earth with my every step. I am someone who will devote herself to creating a happy, comforting, nurturing home. I am someone who will rarely, if ever, abandon another because I cannot bear the thought of someone feeling separate and alone. I am someone who will thoughtfully connect to others as graciously and empathetically as possible.

Are you starting to see how this works? You are so good! The transition occurs when you hit that ideal. There needs to be a massive release. As I said, it is often done through confession, so that you are free to become New. Then you never look back the same way again.

From Glimpse to Expansion. It's not magic. It's how Story works. Every. Single. Time.

Transition Exercise

Please take one of your Ideal traits, and its partner. What MUST happen or has happened, in your life for you to experience the powerful transition that story has to offer?

This is a free write. Please take your time. Be eloquent. Write like you mean it and want to share it.

I'm so proud of you.
YOU ARE INCREDIBLE.
An INSPIRATION to us ALL!
Thank you, Thank you. Thank you.

Transition Time: THE END

The Science of Story

Transition Time: THE END

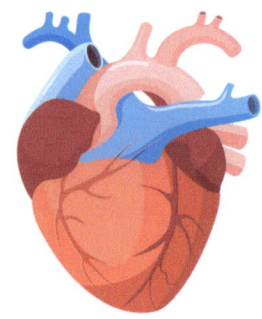

Inside My Story Lab

I think of myself as a Scientist of Story because the role of any scientist is to figure out how things work. In order to properly conduct experiments, a scientist needs a lab to invent, and a process to follow. I offer my process to inspire you to develop yours.

My process evolves but mostly stays the same because it works well for me. I immediately eliminate **anything** that hinders my process. And yes, I am brutal about it. And yes, this has been agonizingly painful at times.

My process follows the seasons.

Winter
I focus on one story (usually Truitt) and I read a lot, more than any other season. Please note winter starts in November in Vermont and stays until April/May so this is nearly half the year for me.

Spring
I create rich, raw, needs-a-lot-of-work material. It shoots out of me so quickly I can't keep up, and don't want to.

Summer
I tend to my garden in the book sense, meaning I edit, and rewrite, and rewrite some more until it goes to my publisher.

Autumn
I enjoy the fruits of my labor. Harvest time. I travel. I connect. I attend and teach at workshops. I pitch and launch new work. I celebrate. And I plan for whatever I'm writing in winter.

My process is a discipline. It is a practice. It requires awareness and contemplation. It gets my full attention and devotion because I have dreams and hopes and desires and promises I've made to myself. I keep my promises.

> **I nurture and honor my process because it is sacred to me.**

The Call to Quiet

I used to write with lots of noise: coffee shops, television, upbeat music, and a variety of other distractions. I had one very important mentor tell me early, early on that I needed to get quiet. I didn't know how. I didn't believe her. I didn't know why it mattered.

Now I do.

I don't do everything in the quiet. But I do a lot. Especially my re-writes. Please note that I said, quiet, not silence. There is a huge difference. Absolute silence is rough. However, when I break free of my need to have a little noise, my writing is better. It's crisper, clearer, stronger. We live in a loud world. It has been become habit to jump our minds from here, to there, to this, and to that, all in a few seconds. To write without the same hops and bounces, I had to form new habits.

Every day starts with coffee and snuggles, then a long walk in the woods or if I'm traveling, on the grounds or streets near my lodging. This is my meditation. It is where I find peace and clarity for the remainder of the day. And yes, if I'm tense or anxious or sad, I take another walk. Nature is my solace, my heaven, my place to surrender. It helps me set the tone of quiet, and that is where I thrive as a writer.

My Notebooks

I create a spiral notebook for each novel, non-fiction book and business I create. I also keep an idea book for each year that is also bound. I take it with me everywhere I go, and I put story ideas or scene ideas or concerns with an idea in that book. It is a macro version of the micros that are individual stories.

All the chapter outlines and thoughts for this book were written down in ink to the paper of a notebook.

I do not allow myself to type one word in the computer until that has happened.

Because if I don't, I write myself into corners.

I don't know what I am creating.

I get lost.

I go off point.

It's ugly. It's frustrating. And it's lazy.

I want to produce the best for you and everyone I interact with.

To do that well, I have to lay it out first. Slowly. And methodically using the tools in the prior chapters. I use every technique I share.

First Drafts: Never Go Back

I write all the way through without stopping, ever. This technique, this gift, was bestowed upon me by the genius writerly whisperer, Anjanette Fennell. She emphatically believes that if you stop and go back to the beginning, even one page back, you lose your flow, momentum, and creativity. You risk the magic. She is right.

I have a novel I have yet to finish, one I have been working on for years, because I keep going backward.

Write your first draft of whatever you are creating until it is done. This includes, but is not limited to, business plans, emails, reports, term papers, products you are making, blogs, articles, webpage content, posts, scripts, plays, letters and journal entries. Keep going. No editing. No going back until the first draft is done. It's not easy to not go back and edit. But you must write forward, never backward.

To relieve my desire/desperate need to change/edit the beginning, I write notes within the manuscript or on the website/program/product page so that I get it out of my head and onto the page.

I write notes directly to myself or my editor in the MS (manuscript), like, "That last bit was crap but I'm moving on." or "The point of that bit is for Truitt to learn that Junius does not know everything even though he does know everything. ONWARD! UGH!" I usually highlight it in bright yellow so I can easily find it and delete it once I have addressed it. And guess what? Usually what I think is crap is pretty good. It was simply a hard and/or emotional section to write so I'm wobbly and fragile and need to vent a moment. (read: confession before new and shiny)

I also put placeholders in the text, highlighted bright yellow. For instance, "Add in party scene details here." Or, "Describe house." Or, "Write love scene."

I write the first draft all the way through and I also edit all the way through each draft. I don't jump around. Ever.

Beginnings

The beginning is not as important as we think. They are highly mutable.

I *always* change the beginning to fulfill the ending.

When I was a "newer" writer, I was told the first three chapters, or first fifty pages matter most. I spent actual years trying to get the first three chapters of my novel, *Becoming Truitt Skye: The City on the Sea,* right and guess what the last thing I did was before it sold? Added chapters to the beginning because my agent had received enough feedback (read: rejection) that it needed to be addressed.

The first three matter once you have created the whole story. It was easy for me to write a couple chapters to the beginning because I knew exactly what would fulfill the reader/publisher. I knew my main character so well and I was happy to do it. The novel sold quickly thereafter.

Second, Third, Sixth, Tenth Drafts

For my "last" draft of any book, I read it out loud. This is the best way to hear it and typos jump off the page. I typically have three "last" drafts before I hand it over to my beloved publisher so they can transform it into a beautiful book for you to enjoy and savor. There will be least three more full edits before my promise to you, and me, is kept and I get to say the words I've been longing to say, "It's Done."

A novel is no different than a business plan, website, sales launch or big report. Every story you create deserves your full attention, does it not? Everything you create is a reflection of you. This is your world. And your stories filling it. Please take the time to go Inside Your Story Lab with every new creation.

Creating Your Story Lab

Developing a process is essential to your success. I'm going to offer some questions to get you thinking about your process. Remember, it is a discipline, a practice and a COMMITMENT to yourself that you will write/create.

Please honor your Process. It is sacred.

What is your most quiet time of the day (please consider creating/writing at this time)?

What room of your home makes you happiest (please consider writing/creating here)?

The Science of Story

What does Winter bring out in you?

Spring?

Summer?

Autumn?

I write best when:

I did the prompts in this journal: (when, where, why):

I am most satisfied when I:

I am going to COMMIT to the following Process for ONE FULL MONTH (and if you want to make this stick, say it out loud to yourself in front of the mirror)

Vetting a Story

Before I begin a story or program or project, I vet it. This process brings everything together. If you have a dream you want fulfilled, vet it, please. This will give it a clear path to arrive directly to you. I would also recommend that you do a Cast of Characters chart when you do your Vetting.

Idea?	

The Science of Story

Who is it about? (this can be the client or boss or publisher you want to attract) Remember this is a version of you, so don't exclude yourself, even it is a fictional character or a client you want to attract. What part of YOU wants to tell this story?	
How does the story/event/program end?	

Vetting a Story

Why? Why does it end that way?	
What is the opposite of the ending?	

The Science of Story

What does life look like for you/the character/client/etc. right now?	
What do you/they believe about yourself/themselves as it relates to this idea?	

Vetting a Story

Who do you/they love? And how will you/they use this to STOP yourself/themselves from reaching the finishing line of this story?	
What matters MOST to you/them?	

The Science of Story

How will what matters most betray you/them?	
What will you/they do?	

Vetting a Story

Who will you/they celebrate with? What will you/they do? Why?	

The Science of Story

Love
Spouse · Parents · Siblings

Challenger

Ex's · "Flirt" Friends · Crushes

Main Character

Bosses · Co-Workers · Teammates

Mentor

Core Group of Friends

Best Friend

Vetting a Story

For the next great idea you have!!

Idea?	
Who is it about? (this can be the client or boss or publisher you want to attract) Remember this is a version of you, so don't exclude yourself, even it is a fictional character or a client you want to attract. What part of YOU wants to tell this story?	

The Science of Story

How does the story/event/program end?	
Why? Why does it end that way?	

Vetting a Story

What is the opposite of the ending?	
What does life look like for you/the character/client/etc. right now?	

What do you/they believe about yourself/themselves as it relates to this idea?	
Who do you/they love? And how will you/they use this to STOP yourself/themselves from reaching the finishing line of this story?	

Vetting a Story

What matters MOST to you/them?	
How will what matters most betray you/them?	

The Science of Story

What will you/they do?	
Who will you/they celebrate with? What you/will they do? Why?	

Vetting a Story

Love

Spouse · Parents · Siblings

Challenger

Ex's · "Flirt" Friends · Crushes

Main Character

Bosses · Co-Workers · Teammates

Mentor

Core Group of Friends

Best Friend

And the next one!

Idea?	
Who is it about? (this can be the client or boss or publisher you want to attract) Remember this is a version of you, so don't exclude yourself, even it is a fictional character or a client you want to attract. What part of YOU wants to tell this story?	

Vetting a Story

How does the story/event/program end?	
Why? Why does it end that way?	

What is the opposite of the ending?	
What does life look like for you/the character/client/etc. right now?	

Vetting a Story

What do you/they believe about yourself/themselves as it relates to this idea?	
Who do you/they love? And how will you/they use this to STOP yourself/themselves from reaching the finishing line of this story?	

The Science of Story

What matters MOST to you/them?	
How will what matters most betray you/them?	

Vetting a Story

What will you/they do?	
Who will you/they celebrate with? What will you/they do? Why?	

The Science of Story

Love
Spouse · Parents · Siblings

Challenger

Ex's · "Flirt" Friends · Crushes

Main Character

Bosses · Co-Workers · Teammates

Mentor

Core Group of Friends

Best Friend

170

Vetting a Story

And the next one!

Idea?	
Who is it about? (this can be the client or boss or publisher you want to attract) Remember this is a version of you, so don't exclude yourself, even it is a fictional character or a client you want to attract. What part of YOU wants to tell this story?	

The Science of Story

How does the story/event/program end?	
Why? Why does it end that way?	

Vetting a Story

What is the opposite of the ending?	
What does life look like for you/the character/client/etc. right now?	

What do you/they believe about yourself/themselves as it relates to this idea?	
Who do you/they love? And how will you/they use this to STOP yourself/themselves from reaching the finishing line of this story?	

Vetting a Story

What matters MOST to you/them?	
How will what matters most betray you/them?	

The Science of Story

What will you/they do?	
Who will you/they celebrate with? What will you/they do? Why?	

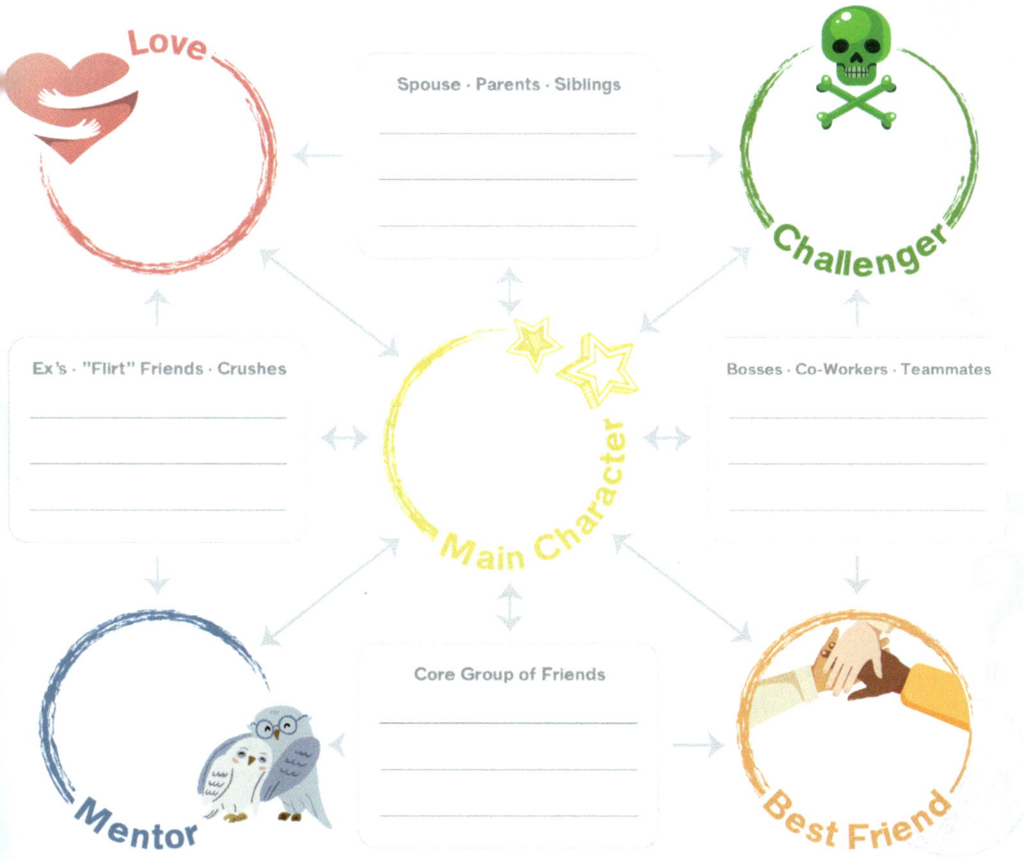

Vetting a Story

Love

Spouse · Parents · Siblings
_____
_____
_____

Challenger

Ex's · "Flirt" Friends · Crushes
_____
_____
_____

Main Character

Bosses · Co-Workers · Teammates
_____
_____
_____

Mentor

Core Group of Friends
_____
_____
_____

Best Friend

From Here Forward

Take your time. Rush. Think. Don't think. Go off track. **Stay on track.** Pace yourself. Go there. Be there. Become there. Creating is a strange place between going fast and taking your sweet time. Between finding meaning and letting meaning find you. Story as Dance.

I know you can be your dream. I know you can sit still and let your creation flow. I know you have the desire to complete whatever you start. I know you have the skills. I know you possess the stick-to-it-ness this takes. You have everything you need within you. Let yourself get distracted when your story/life doesn't hold your attention. Close your notebook, computer program, or discussion when you start to feel sleepy or bored. It can wait. You won't forget. It will marinate and get better with each passing day. Live your story all the way.

Trust yourself fully. Trust that you know what you're doing because you absolutely do. Trust that what you don't know, you will discover. Schedule time with a writing coach or life coach or counselor or massage therapist or yogi, whomever you need to help you trust yourself as fully as you possibly can.

You are the voice of you.

E v e r y t h i n g you are matters.

You are a one of a kind, unique, beautiful, desirable soul.

Please let who you are shine. One creation at a time.

I wish you laughter, love, peace, and above all, hundreds of wondrous stories along the way.

My heart to yours,

adrea.

Acknowledgements

I thank each and every one of you for being here and wanting to know more about Story. The more I share it, the more I learn it. I cherish the exhilaration of seeing Story in action. I have had the addictive pleasure of seeing my clients and friends relax and release their anxiety/terror when I remind them why they are acting this way or that way. "Of course this crap is happening. It has to! That's how Story works! Triumph is guaranteed, remember?" I hope you will join me for a workshop someday. I would love that!

Thank you, Cassandra Neece, of the Dharma Collective, for designing the outside (cover) and inside of this book and journal and all the marketing materials, including creating the infographics for The Guts of Story and their beautiful matching icons. You are a genius. Cass also is the creator my website, www.adreapeters.com. She truly is the magic in all of my creations. Thank you, Cass. THANK YOU.

To the ridiculously talented and generous, Amber Lilyestrom, who saw this book and all that is now becoming in my life wayyyyy before I did. My mentor, my dear friend, and a Super Heroine of the very best kind. You know how much I love you, right?

To the divine, Michelle Lowbridge, thank you for letting me share and vet this material with you!

To Anj, always in my mind and heart and soul with every word I type. Thank you. You are exceptional and incomparable.

To my brilliant publisher, Karen McDermott. Thank you for loving the way I think and reminding me that people will connect with my offerings. Your support and encouragement means the absolute world to me.

To the entire Making Magic Happen team, thank you all! You are so talented! I am in awe of all that you do for the authors of this community, Thank you.

To the teachers that helped me form an emotional vocabulary and love of self, Abraham-Hicks, Wayne Dyer, Caroline Myss and Louise Hay, Thank you.

To all my writing mentors, I would never be writing this without you. Thank you, Beth Gaeddert, Del Langbauer, Mollie Gregory, Carolyn See, Bob Bausch, Tim Perrin, Lisa Catherine Findlay, John Truby, John Schimmel, Dara Marks, Robert Menna, Ann Hood, Francesco Sedita, Don Maass, Anjanette Fennell and the faculty at Seton Hill University, where I earned my Master of Arts in Fiction Writing in 2008.

About the Author

Adrea is a novelist and screenwriter who dramatically changed her life through her writing. "There is a fundamental correlation between our stories and who we are in our daily life. We create what we imagine. We are not separate from our souls. We are not apart from our dreams. In order to know this, to truly live this, we must walk our talk and love ourselves completely." Adrea knows how challenging and ambitious that is, yet, it does not stop her from trying every single day.

In addition to her own writing and teaching, she has mentored aspiring fiction, memoir and screenwriters. She has been teaching these techniques for over a decade. She also runs a successful business consulting practice.

Adrea graduated Valedictorian from the University of Colorado at Boulder with a Bachelor of Science in News Writing, earned a certificate as a Holistic Nutrition Educator from Bauman College, and was awarded her Master of Arts for Fiction Writing from Seton Hill University.

Her novel series, *Becoming Truitt Skye*, takes up most of her writing life these days. She has also published several inspirational alphabet books with Teffanie Thompson of *pictureless books* and is a contributing writer to the award-winning textbook for writers, *Many Genres, One Craft*.

Prior to her commitment to her novels and screenplays, she wrote articles for newspapers and magazines, including a contributing role with 831magazine where she wrote food articles, interviewed food experts, and had a standing nutrition humor column counting down "Top 8" for a variety of wellness topics, including *8 Ways to Cure a Hangover, Top 8 Excuses for Eating Really Bad Food* and *8 Ways to Cure Insomnia*.

The Science of Story includes a *Companion Journal* and a workshop offering.

Please visit her website for more details.
www.adreapeters.com

About the Designer

Cassandra (Cass) is a graphic designer, marketing maven and owner of The Dharma Collective, a branding studio focused on serving passionate entrepreneurs. She works with heart-centered visionaries to bring their vision to life, one that embodies their mission and engages their ideal client with a holistic strategy. She believes in raising the collective consciousness by encouraging others to connect with their innate gifts and talents so they can create that which they wish to see in our world.

When she's not co-creating with her clients, you can find Cass deepening her yoga practice, adventuring in the great outdoors and spending quality time with the ones she loves.

For more information about Cass and The Dharma Collective visit
www.thedharmacollective.com

www.ingramcontent.com/pod-product-compliance
Lightning Source LLC
Chambersburg PA
CBHW041431010526
44107CB00046B/1572